www.mascotbooks.com

The ABCs of the Army

For more information, please contact:
Mascot Books
560 Herndon Parkway #120
Herndon, VA 20170
info@mascotbooks.com

Library of Congress Control Number: 2016910705

CPSIA Code: PRT0916A
ISBN-13: 978-1-63177-656-4

Printed in the United States

The ABCs
of the Army

by Maria Cordova

Illustrated by Jeff Mora

To my husband, "Doc" Chris Cordova, and my girls, Kylee and Ella—my strength, my Army.

For the Gold Star families I had the pleasure of meeting. You will always be in my heart forever. Thank you for allowing me to be a part of your family.

To the Hoover family.

Aa

is for Airborne. In the Army, soldiers learn to jump out of planes and land safely! They can even "earn their wings" to become a paratrooper after completing the three-week course at Fort Benning, Georgia.

Bb

is for Basic Combat Training (BCT). BCT is a ten-week course to turn civilians into soldiers. It's the first experience of real Army life.

Cc

is for Combat Medic. The combat medic is the first person to provide medical care to the wounded on the battlefield.

Dd

is for Deployment. Deployments occur when your soldier is in another country far away from family for a few months, sometimes a year, to defend our nation and keep us safe.

Ee

Duties

✓ Formation at 0600

✓ PT at 0630

✓ Meeting with NCOIC 0900

✓ Motorpool at 1000

✓ Clean weapons at 1300

✓ Training at 1500

GRIFFIN   US ARMY

is for Enlisted. An enlisted soldier will be tasked for everyday duties and missions. The enlisted are the working men of the Army. Enlisted soldiers always complete their tasks!

Ff

FRGs consist of amazing spouses, volunteers, and a command group who support each other by organizing events and meetings during deployments. They will be your battle buddies during each deployment.

Gg

is for Guard Duty. Soldiers guard and protect their post until their guard duty is complete.

Hh

HOOAH!

is for Hooah! This is an animated yell meaning "yes" and never "no." Hooah!

Ii

is for Infantry or the "Queen of the Battle." The infantryman is an expert in weapons and individual maneuvers and tactics.

Jj

is for the month of June. The Army's birthday is June 14, 1775. Don't forget to eat cake and ice cream to celebrate!

America's Army:
Strength of the Nation

JUICE!

Kk

8 LAPS = 3.2 KM (2 MILES)

is for Kilometer. Exactly 3.2 kilometers (or 2 miles) + push-ups + sit-ups = The Army physical fitness test. If one lap around the track field equals .40 kilometers (1/4 mile), how many laps does it take to equal 3.2 kilometers (2 miles). What do you think?

L l

is for Land Navigation. If you can read a map and use a compass, you can find your destination. If not, you might get lost. Land navigation is a useful skill, especially in Ranger School.

Mm

COFFEE

CRACKERS

SUGAR

CHIPS

Meal, Spaghetti with Sauce

COOKIE

is for Meals, Ready-to-Eat (MRE). An MRE is like a TV dinner that can be eaten without heating. Sometimes the taste isn't great, but an MRE does its job and provides energy for hungry soldiers! Did you know an MRE is good for up to three and a half years if stored properly at 80 degrees Fahrenheit?

Nn

is for National Guard. The National Guard is called upon when national disasters occur in the United States. The National Guard conducts training drills one weekend a month, and annual two-week training sessions.

Oo

is for Old Guard. The Old Guard is known today as the 3rd U.S. Infantry Regiment and is the formal ceremonial unit and escort to the President of the United States. Have you ever visited Arlington National Cemetery in Virginia? There you'll always find a soldier marching back and forth at the Tomb of the Unknown Soldier. This soldier is known as a Tomb Sentinel and is part of the Old Guard.

Pp

is for Permanent Change of Station (PCS). PCS, or moving orders, are orders received from your branch manager informing you and your family where you will be moving next. A soldier typically moves every three years.

Qq

"Be there in 5 minutes."

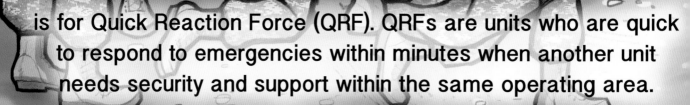

is for Quick Reaction Force (QRF). QRFs are units who are quick to respond to emergencies within minutes when another unit needs security and support within the same operating area.

Rr

RANGER

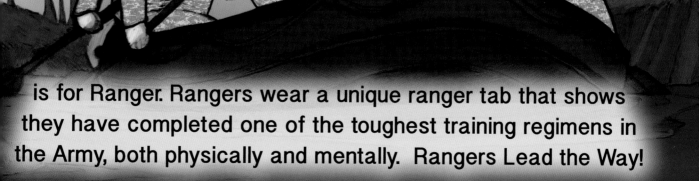

is for Ranger. Rangers wear a unique ranger tab that shows they have completed one of the toughest training regimens in the Army, both physically and mentally. Rangers Lead the Way!

S s

is for Song. Yes, the Army has an official Army song, which is sung loud and proud at all Army events and ceremonies. "The Army Goes Rolling Along!"

"THE ARMY GOES ROLLING ALONG!"

March along, sing our song, with the Army of the free.
Count the brave, count the true, who have fought to victory.
We're the Army and proud of our name!
We're the Army and proudly proclaim:

First to fight for the right,
And to build the Nation's might,
And The Army Goes Rolling Along.
Proud of all we have done,
Fighting till the battle's won,
And the Army Goes Rolling Along.

Then it's hi! hi! hey! The Army's on its way.
Count off the cadence loud and strong; For where'er we go,

You will always know
That The Army Goes Rolling Along.

Tt

is for "This We'll Defend," which has been the official Army motto since its birth in 1775! Did you know the official Army emblem was not established until January 29, 1974 by the Secretary of the Army?

THIS WE'LL DEFEND

DEPARTMENT OF THE ARMY · UNITED STATES OF AMERICA

1775

Uu

is for United States Military Academy (USMA) at West Point, New York. Cadets are commissioned as second lieutenants in the United States Army upon completing four years of college.

Vv

is for Veteran. A veteran is someone who has served honorably in the United States Armed Forces. Veterans have their own federal holiday, known as Veteran's Day, on November 11th. If you know a veteran or one is reading this book to you, always make sure to thank them, too!

Ww

is for Woobie. Most soldiers would agree this is the best invention ever! What exactly is it? It's a huge, camouflage soft blanket that will keep you warm and make great forts, too! I bet every soldier has one in his living room.

Xx

X EXCHANGE

is for eXchange, as in the Post Exchange (PX), which is similar to Target and Walmart and not a store to exchange items.

Yy

is for Yellow Ribbon. The yellow ribbon symbolizes troop support and hope during a time of war. Sometimes you will see yellow ribbons tied around a tree or a bumper sticker on a car.

is for Zonk! When the First Sergeant yells "Zonk!" during formation, this means "No physical fitness today! Go home!" Every soldier has a smile on their face.

About the Author

Maria currently lives in Colorado Springs, Colorado, with her active duty Army husband Chris and two beautiful girls, Kylee and Ella. Although Maria is married to the Army, she is a Navy brat and cheers for both Army and Navy. Go Army! Go Navy!